Rainforest animals near and far—
Can you tell me where they are?

Keel-Billed Toucan

Who's **behind** the leaves?

Jaguar

Who's up a tree?

Golden Lion Tamarin

Who's between the leaves
and hard to see?

Casque-Headed Iguana

Inside or outside?

Hyacinth Macaws

Above or below?

Chameleons

Who's in the middle of the row?

Poison-Arrow Frogs

Who's swimming under water?

Crocodile

Margay Ca

Who's jumping
through the air?

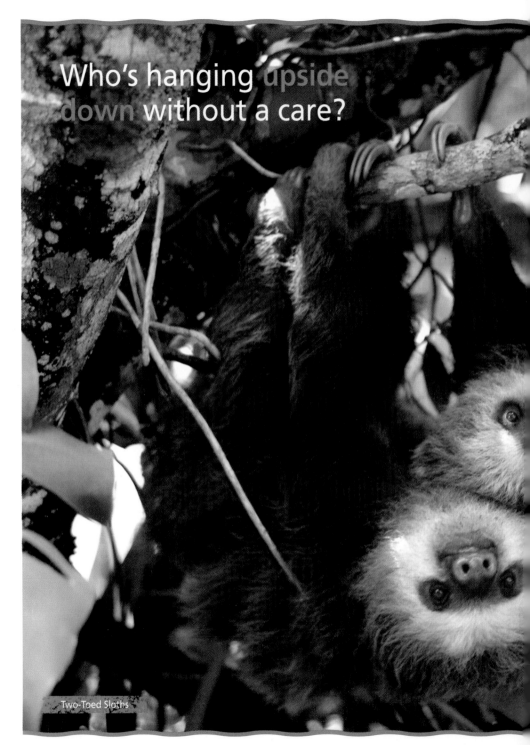

Who's hanging upside down without a care?

Two-Toed Sloths

Who's on **top**
going for a ride?

Squirrel Monkey

Bearded Saki

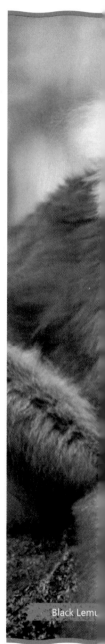

Black Lemur

Who's sitting together
side by side?

In the air,

Green Violetear

in the water,

Brazilian Tapir

and on the ground—

Capybaras

Rainforest animals are all around!

Emerald Tree Boa